THE TAJ MAHAL

Karen Bush Gibson

PURPLE TOAD PUBLISHING

Printing 1 2 3 4 5 6 7 8 9

BUILDING ON A DREAM

Big Ben
The Eiffel Tower
The Space Needle
The Statue of Liberty
The Sydney Opera House
The Taj Mahal

Publisher's Cataloging-in-Publication Data
Gibson, Karen.
Taj Mahal / written by Karen Gibson.
p. cm.
Includes bibliographic references, glossary, and index.
ISBN 9781624692116
1. Taj Mahal (Agra, India)—Juvenile literature. 2. Architecture—Vocational guidance—Juvenile literature. I. Series: Building on a Dream.
NA2555 2017
507.8

Library of Congress Control Number: 2016937176

eBook ISBN: 9781624692123

ABOUT THE AUTHOR: Karen Bush Gibson enjoys discovering other cultures, civilizations, and people. She has written over 30 books for children that bring history alive. Gibson lives with her family in Oklahoma.

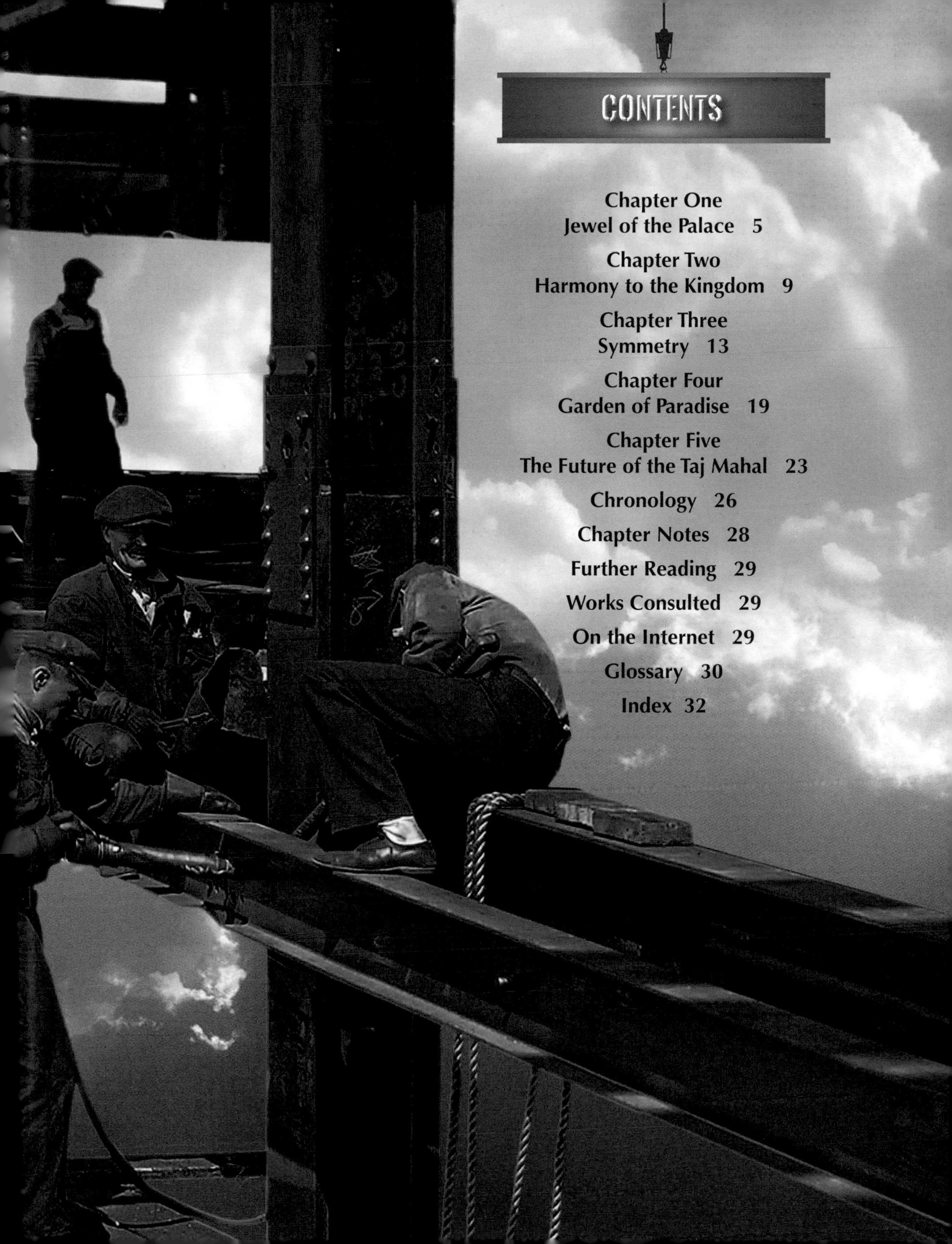

CONTENTS

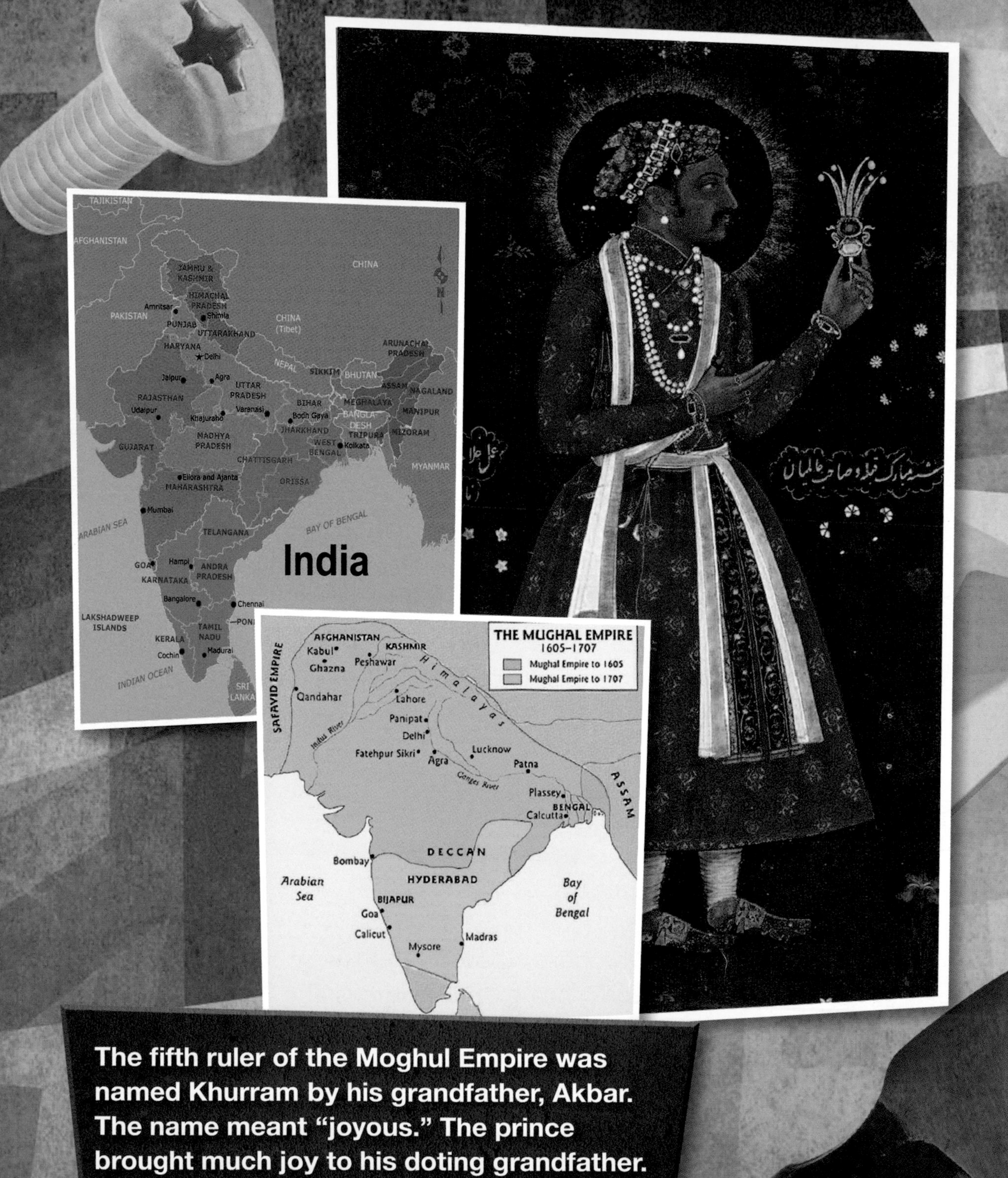

The fifth ruler of the Moghul Empire was named Khurram by his grandfather, Akbar. The name meant "joyous." The prince brought much joy to his doting grandfather.

CHAPTER 1

Jewel of the Palace

In India, a grand palace stands on a riverbank. Bright with lush gardens and colorful jewels on white stones, it is fit for royalty. But this Wonder of the World was not built for the living. Instead, it is a tomb for a beloved queen, wife, and mother.

India has a long history of great civilizations. Mark Twain once called this South Asian country "the cradle of the human race"—meaning civilization was born here.[1]

One of the greatest dynasties to rule India was the Mughal Dynasty. These Persian descendants of Genghis Khan ruled South Asia from 1526 to 1858. Mughals wanted to expand their territory and leave their mark on the world. They succeeded at both. The Mughal Empire included today's India plus Afghanistan, Pakistan, Nepal, Bhutan, and Bangladesh.

One Mughal leader was more interested in creating wonderful buildings than conquering. As a boy, he was known as Khurram. From a young age, Prince Khurram had a passion for building. His grandfather, Akbar the Great, had built the famous Red Fort in the city of Agra. It contained palaces and living quarters for the royal family. When Khurram was 16 years old, he built his own home in the complex.

Khurram also redesigned many of the buildings and created the Peacock Throne. This massive golden chair had a domed roof supported by twelve columns. It featured the finest jewels in the empire, including diamonds, rubies, and pearls.

By age 25, Khurram had a string of military successes that had expanded the southern border of the kingdom. His father, Emperor

Jahangir, was so impressed that he gave his favorite son a new title—Shah Jahan Bahadur. This Persian title means "King of the World."[2] Along with the title came the throne. Shah Jahan became the fifth Mughal ruler when his father died in 1627.

The period between the rule of Akbar the Great and that of his grandson, Shah Jahan, is often called the Classic Period of the Mughal Empire. Although Mughals were Muslim, the people they ruled were largely Hindu. Rulers during the Classic Period wanted religious harmony, and they encouraged Hindus and Muslims to marry. It was a time of great wealth, and both trade and the arts grew.

An artist's idea of Mumtaz Mahal's beauty. Shah Jahan's true love ordered a riverside garden to be built in the city of Agra.

Mughal leaders were expected to have many wives. Shah Jahan had three, but found true love with his second wife, Arjumand Banu Begum. He called her Mumtaz Mahal, meaning "Jewel of the Palace."[3] Although painting portraits of Mughal women was

The Taj Mahal is named after Mumtaz Mahal. Its name means "Crown Palace," or "Crown of the Palace."

prohibited, poets and court reporters commented on her beauty, compassion, and devotion to her husband.

Mumtaz traveled everywhere with her husband, even when he went to battle. In 1631, while pregnant with her fourteenth child, she traveled with Shah Jahan to southern India. Soon after giving birth to a daughter, Mumtaz knew she was dying. She asked her husband to build her a beautiful tomb that he could visit on the anniversary of her death.

Shah Jahan was heartbroken at the loss of his beloved. He traded his colorful jeweled robes for the white robes Hindus wear for mourning. His hair and beard seemed to grow gray overnight. But he did not forget his promise. Shah Jahan would honor his wife by building her the most magnificent tomb in the world, the Taj Mahal.

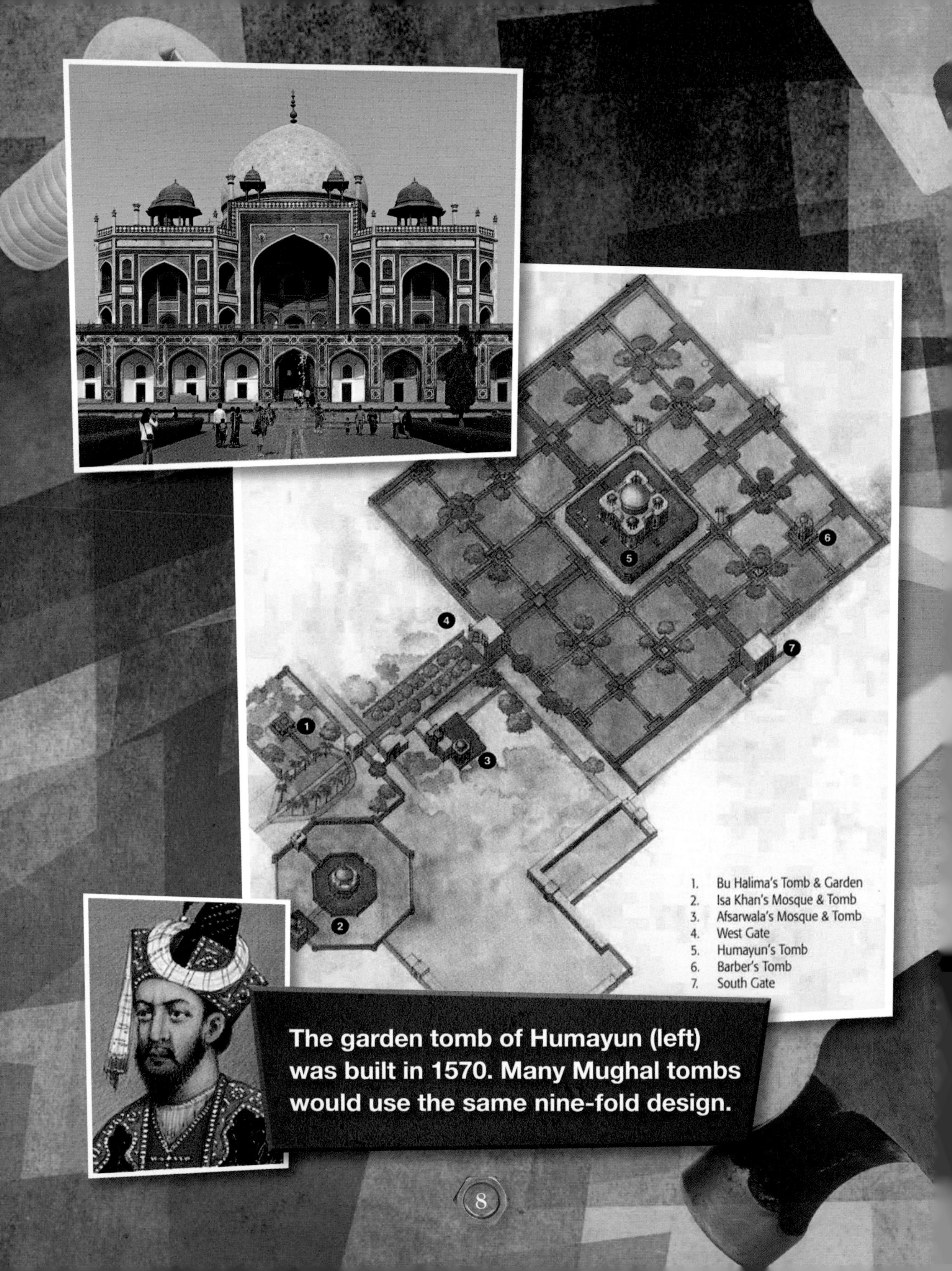

The garden tomb of Humayun (left) was built in 1570. Many Mughal tombs would use the same nine-fold design.

CHAPTER 2

Harmony to the Kingdom

Honoring the dead is an ancient custom found in many cultures. For example, the pyramids of Egypt are tombs that were built for kings and queens. In Mughal culture, paying tribute to the dead was very important, especially among the ruling class. Instead of tombstones, complex buildings called mausoleums were built to house the remains of the dead. Special celebrations were held each year on the anniversary of a death.

The Mughal nobility enjoyed beautiful garden areas. The death of Humayun, Shah Jahan's great-grandfather, led to the first tomb set in a garden.[1] Humayun's tomb combined the beauty of gardens with fancy sandstone buildings. The tombs brought together Hindu, Islamic, and Persian features. Like the people and religions of the time, Indo-Islamic architecture was a harmony of many styles.

Mughal architecture for tombs and gardens was often based on a nine-fold plan that created an octagon. In the nine-fold plan, four intersecting lines divide the building area into eight parts. A domed chamber in the center is the ninth part. The eight sides of an octagon represent the eight divisions of the holy book of Islam, the Quran. A central axis runs through the main dome, and each side balances the other. This symmetry symbolizes the power of a ruler bringing harmony to the kingdom.

These ideas are part of the design of the Taj Mahal. But the Taj Mahal expanded Indo-Islamic architecture with curved forms and different styles of columns. Air shafts were also included in the design.

The white marble tombs of Mumtaz Mahal and Shah Jahan are cenotaphs, meaning "likenesses of the tombs." The bodies are actually entombed in a chamber below the cenotaphs.

Another difference was the building materials. Although sandstone would be used, parts of the Taj Mahal would be cloaked in white marble. White marble would focus attention on the tomb of Mumtaz. In Hinduism, colors have different meanings. White belonged to the ruling or priestly caste. Red represented warriors.

Who was the actual architect of the Taj Mahal? No one knows for certain. When it was built, credit was given to its patron—Shah Jahan. He had already shown a talent for building, and he wanted to make the Taj Mahal an earthbound copy of Mumtaz Mahal's home in the afterlife. It is almost certain that the ruler played an important part in the design of the building. But he also had help from the greatest builders, designers, and artisans of the time.[2]

Official Mughal histories list the names of 37 men involved in the design and construction of the Taj Mahal. Ismail Afandi (Ismail Khan) of the Ottoman Empire was known for building domes. He most likely designed the main dome, which is often called an onion dome due to its shape. Many Hindu craftsmen and laborers turned the plans into real structures.

Location was essential to the design. Water was precious in the dry plains of India. The kingdom of Agra was built on the banks of the Yamuna River. A spectacular tomb for Mumtaz would certainly include gardens, which were prized in the Muslim religion. The Quran speaks of the "garden of paradise." The Taj Mahal, then, would also be built on the banks of the Yamuna River.

The twelfth-century Tatar tribe were probably the first people to make onion domes. The Tatars later became part of the Mughal Empire.

Oxen are often used to help maintain the grounds around the Taj Mahal.

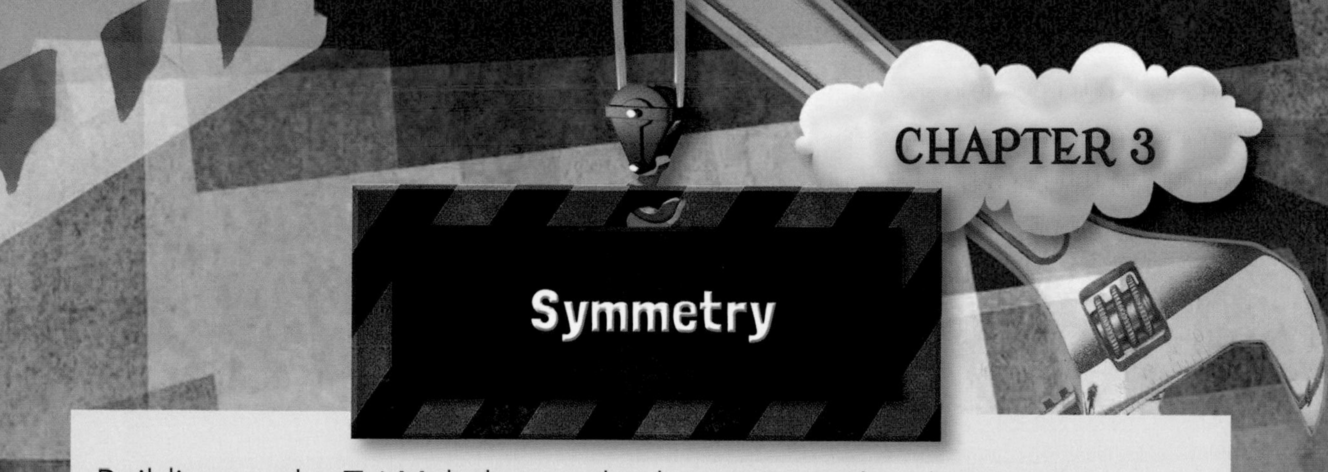

CHAPTER 3

Symmetry

Building on the Taj Mahal started in late 1631 and took 22 years to complete. No one knows how many laborers worked on it. There may have been 5,000 or as many as 20,000.[1] The workers lived south of the complex in houses that Shah Jahan had built for them. Soon it was a busy neighborhood, with shops and vendors lining the street.

Structures the size of the Taj Mahal are built on strong foundations. Without one, a building can literally fall apart. At 12,000 tons and located on a riverbank, the Taj would need an especially strong foundation.

Have you ever been to a beach? The sandy ground constantly shifts as water moves in and out. The soil where the Taj Mahal was to be built was like beach sand—unstable. Without special modifications, the building would be unstable as well.

Building engineers dug deep wells into the riverbank until they hit solid ground. Workers packed the wells with rocks and concrete. A platform 23 feet thick was laid across these concrete pillars. The platform, or plinth, was made of sandstone. A marble platform in a black-and-white checkerboard pattern was laid on top of the plinth.

The mausoleum had double walls. The inner wall was made from small 7-inch-square bricks. Concrete and iron clamps held the bricks together.

The outer walls were sandstone blocks. Sandstone, the material used to build the Agra Red Fort, was easy to find in northern India. To move it, a 10-mile-long raised road was built. Carts pulled by oxen hauled the sandstone from the quarry to the building site.

At the construction site, masons used hammers and chisels to cut the sandstone to size. They smoothed the edges and faces of the rocks by rubbing them with grit, the sandpaper of the 1600s. They started with coarse grit and moved to finer grit.

Not all of the Taj Mahal was sandstone. The main mausoleum and corner towers were faced in white marble. White marble was more difficult to shape than sandstone. Mined approximately 200 miles away in Persia, it is said that 1,000 elephants were used to move these stones to Agra.

Four narrow marble towers stand at the corners of the mausoleum. Their domes copy the main onion dome.

Shah Jahan hired every stonemason he could find who had experience working with marble. As with sandstone, the white marble had to be cut to size and sanded smooth. Although heavier, marble was more fragile to work with. A single wrong move could chip or crack a piece of marble.

One building dilemma was how to support a round dome on a square building. The answer was to create an arch where two walls meet. This turned

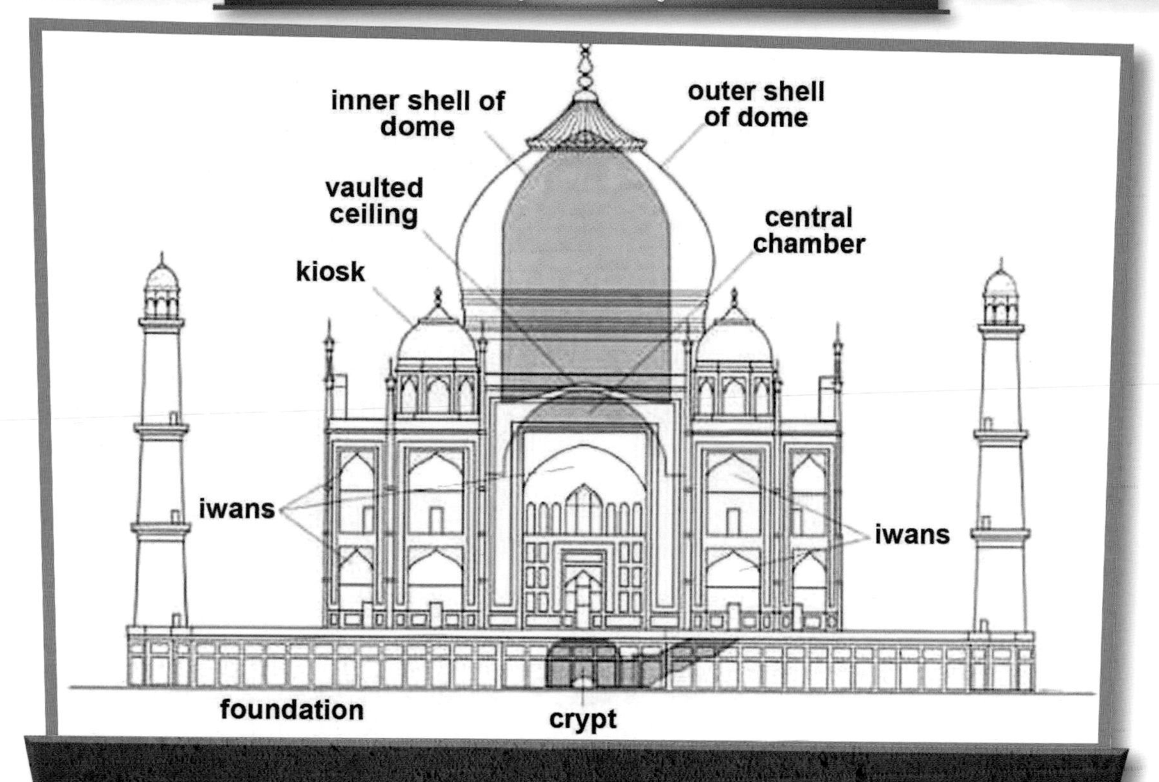

The main dome has an inner and outer shell. This is called a double-skinned dome.

a four-sided building into an eight-sided or octagonal building. These walls could support a dome as wide as 58 feet across.[2]

The mausoleum would be 240 feet tall. Builders had to find a way to get the materials to the top of the walls. Like the ancient Egyptians had done when constructing the pyramids, the workers built scaffolding and ramps. Bamboo or other wood was used for most scaffolding in India. At the Taj Mahal, brick scaffolding was used instead. The reasons can only be guessed. Did the brick scaffolding provide more support in moving materials? Or did Shah Jahan wish to hide the work of the main mausoleum from curious eyes?

Eight areas surround the octagonal tomb. Four corner rooms are topped by four smaller domes. Unlike the main dome, the smaller

The Taj Mahal has a terrace that looks out on the Yamuna River.

domes have circular balconies known as kiosks. Each side of the main building features a wide recessed arch 108 feet high. Smaller arches frame the central arch. A 984-foot-long terrace was built on the north side, overlooking the river.

All of the Taj columns are the same style. Called a Shahjahani column, each has a base of four arched panels and a shaft with many faces. The size and detail of the columns vary. In the outer areas, the columns are simple. The columns in the mausoleum are more decorative.[3]

By 1643, the mausoleum was completed. Shah Jahan was told it would take five years to take down the brick scaffolding. Instead, the emperor told the people that whoever took down the bricks could keep them. The scaffolding came down overnight.[4]

In June that year, the Taj Mahal held its first death anniversary. But there was still more to do.

The Taj Mahal is not a single building. It is a 42-acre complex with a garden and other buildings. There is a guesthouse or assembly hall, designed for visitors who came on the death anniversary. A mosque near the main building is used for prayer. The floor is inlaid with 569 prayer rugs, outlined in black marble. A red-orange sunburst design decorates the ceiling.

A wall surrounds the complex, with four slender white marble towers stretching more than 137 feet high at each corner of the plinth. These towers, called minarets, are used to call Muslims to prayer.

The first level of each minaret is the same height as the first level of the mausoleum. The minaret domes align with the widest part of the mausoleum's main dome.

The Taj Mahal has four gates: north, east, south, and west. Shah Jahan and the royal family used the north gate. It is made of red sandstone inlaid with white marble and decorated with plants, flowers, and words from the Quran.

The south gate is topped with a series of domes and features recessed arches. The most photographed view of the Taj Mahal—which shows the garden that leads to the main mausoleum—is seen from the south gate.

No matter which gate visitors use to enter the grounds, the views of the Taj Mahal are spectacular.

Italian artists were hired to teach Mughal people how to cut the precious stones that decorate the Taj Mahal.

Garden of Paradise

Attention to detail continued inside the buildings and in the gardens. Delicate inlays decorate inside and out. Inlay work is an art form that uses materials of different colors and textures. Artists put pieces of these materials together to create a picture. The finished piece is polished smooth.

In the mausoleum, the inlays look like flowers and plants. They are made with black and yellow marble, jade, jasper, and other types of colorful stones. So many lilies, lotus leaves, and other flowers were created in the marble and sandstone that the interior looks like another garden. Even the ceiling is etched with flowers.

The main dome is about 66 feet smaller inside than outside. According to the Quran, tombs should be open to the sky. Lowering the inner dome symbolized the sky.

Another highly valued artist was the calligrapher. In today's culture, we usually don't think of handwriting as art. But in Mughal times, calligraphers were in great demand in royal circles. The Persian calligrapher Amahat Khan was the only craftsman allowed to sign his work at the Taj Mahal. Black marble words in the tomb are from the Quran.

The garden was the last part to be built. Completed in 1653, it too is symmetrical. It features a rectangular pool that runs south from the mausoleum. The water reflects the mausoleum. Another pool crosses the main channel, dividing the garden into four equal squares. Originally, each square contained 16 flowerbeds with 400 plants.

At the Taj Mahal, aqueducts supplied water to the fountains and gardens.

An aqueduct carried water from the river to the east side of the main building. From there, pulleys were used to lift the water in animal hides and dump it into storage tanks at the top of a wall.[1] Water flowed from the tanks into channels, and then into pipes that filled the pools and fed the fountains. The water also irrigated the gardens, creating a lush paradise.

In his later years, Shah Jahan became ill. Four of his sons fought over the throne. The third son, Aurangzeb, won. He then locked his father in his private chambers. From his window in an octagonal tower, Shah Jahan gazed upon the Taj Mahal for the next eight years. When he died in 1666, Aurangzeb buried him next to his true love.

It was rumored that Shah Jahan had been building his own tomb in black marble on the other side of the river.[2] Archeologists have found no evidence of black marble or any tomb there.

It appears that Shah Jahan did not plan to be buried next to Mumtaz. To build another tomb next to hers would have changed the balance of his original design. The tombs of Mumtaz and Shah Jahan are side by side under the mausoleum. They were placed facing west to Mecca, the holy city of Islam.

Empty tombs of the royal couple are directly above the real tombs. The empty tombs at floor level are known as cenotaphs. On top of Mumtaz's cenotaph is a carving of a writing slate. On Shah Jahan's cenotaph sits a carved inkwell. The carvings symbolize a man writing his desires on a woman's heart.

Marble lattice surrounds the cenotaphs of Shah Jahan and Mumtaz Mahal. Each lattice wall was carved from a single piece of marble.

The Taj Mahal cost almost a billion dollars to build in today's money. The cost led to famine, as food was taken away to feed the workers.

CHAPTER 5

The Future of the Taj Mahal

Building the Taj Mahal cost a lot of money. Emperor Aurangzeb was disgusted by this extravagance. He also changed many of Shah Jahan's laws. The Mughal dynasty began to crumble under his leadership.

European powers were interested in India. By 1757, Great Britain's East India Company was fighting local leaders for control of land and trade. About 100 years later, India became a British colony. It remained under British rule until 1947, when India became independent again.

When the empire fell, thieves stole precious gems and metals from the Taj Mahal. Some people chiseled pieces of stone to take home as souvenirs. Lord Curzon, the British head of state in India from 1899 to 1905, ordered repairs to the Taj. Finally, in 1908, it opened to the public. In 1983, "the jewel of Muslim art in India" was named a UNESCO World Heritage Site.[1]

Although many visitors enter the grounds through the east or west gate, the view from the south gate is perhaps the most stunning. From the south, the mausoleum changes color with the light of the day. In the early morning hours, it takes on a soft pink color before turning milky white. In the full sun of the day, it sparkles. And in the moonlight, it can range from silver to the winking colors of a pearl.[2]

The summer solstice reveals even more about the Taj Mahal. On the day of the year when the sun is highest in the sky (usually around June 21), the Taj Mahal is aligned with the path of the sun. An Italian

As the sun moves across the sky, the colors of the Taj Mahal change.

physics professor, Amelia Carolina Sparavigna, discovered this perfect alignment. A visitor standing in the garden where the two pools cross can watch the sun rise over the northeast pavilion. It sets over the northwest pavilion on the other side. The mausoleum and minarets are centered between those pavilions. As it rises and sets, the sun appears to frame the Taj Mahal.[3]

To view the inside of the mausoleum, visitors must take off their shoes or wear booties over them to protect the marble floor. Photography is not allowed inside the buildings. Flashlights are provided to help visitors see the artwork in the shadows of the tomb.

Since 1947, the Archaeological Survey of India has been in charge of protecting the Taj Mahal. Craftsmen whose families have worked on the Taj Mahal since the beginning work closely with the Survey. They replace marble and sandstone slabs as needed. They cut the stone with chisels and polish it by hand, much like their ancestors did.

The condition of the buildings is closely watched. Pollution can stain the marble and make it decay. In 1996, India's Supreme Court ruled that pollution should be restricted around the Taj Mahal. The ruling banned coal industries from the Taj Trapezium Zone (TTZ), an area of about 6,500 square miles around the complex. Fuel-burning cars are not allowed within about one-third mile of the gates.[4]

In 2016, pollution on the Yamuna River allowed great swarms of insects to thrive. The insects rested on the walls of the Taj Mahal, turning them green. Scrubbing the walls created more damage.

As the Taj Mahal approaches 400 years old, it remains one of the most famous buildings in the world. Eight million people visit it each year.

Architecture has been called the marriage of engineering and art. One look at the Taj Mahal proves that truth.

The Taj Majal is the most visited tourist place in India.

1556 Akbar is considered the greatest of the Mughal emperors and is often referred to as Akbar the Great.

1565 Akbar builds the Red Fort at Agra.

1592 Prince Khurram (Shah Jahan), the son of Jahangir, is born.

1593 Arjumand Banu Begum (Mumtaz Mahal) in born.

1605 Akbar dies and is succeeded by his son, Jahangir.

1612 Shah Jahan and Mumtaz Mahal marry.

1613 Emperor Jahangir allows the British East India Company to trade goods with India.

1627 Shah Jahan becomes the fifth Mughal emperor.

1631 Mumtaz Mahal dies; Shah Jahan begins building the Taj Mahal.

1643 The mausoleum of the Taj Mahal is complete.

1644 The British East India Company builds a fort in India.

1653 The rest of the Taj Mahal—buildings, garden, and gates—is finished.

1658 Aurangzeb, a son of Shah Jahan, defeats his brothers for the Mughal throne. He imprisons his father in the Red Fort.

1666 Shah Jahan dies. Emperor Aurangzeb decides to bury him in the Taj Mahal, next to Mumtaz Mahal.

1707 Aurangzeb dies, and the Mughal Empire falls.

1757 The British East India Company controls India through trade and military action.

1858 India becomes a British colony. The last Mughal emperor, Bahadur Shah II, is exiled to Burma.

1908 The Taj Mahal is restored.

1947 India becomes an independent country.

1983 The Taj Mahal is named a UNESCO World Heritage Site.

2015 Approximately 8 million people visit the Taj Mahal.

2016 Swarms of insects land on the walls of the Taj Mahal, turning them green. Efforts to scrub the walls cause more damage.

Sunset at the Taj Mahal

Chapter 1. Jewel of the Palace

1. Mark Twain, *Following the Equator: A Journey Around the World* (Arlington Heights, IL: Twain Press, reissued 2011), p. 186
2. "Taj Mahal" National Geographic Traveler http://travel.nationalgeographic.com/travel/world-heritage/taj-mahal
3. Pamela D. Toler, *Mankind: The Story of All of Us* (New York: Running Press, 2012), p. 256.

Chapter 2. Harmony to the Kingdom

1. "Humayun's Tomb, New Delhi," World Heritage Convention, http://whc.unesco.org/en/list/232
2. Barry Stoner (producer), *Treasures of the World: Taj Mahal,* "Who Designed the Taj Mahal?" 1999. http://www.pbs.org/treasuresoftheworld/taj_mahal/tlevel_2/t3build_design.html

Chapter 3. Symmetry

1. Steven Carroll (director), *Taj Mahal* (A & E Television Networks: Distributed by New Video, 2010).
2. *Deconstructing History: Taj Mahal* (History Channel) http://tinyurl.com/hgxvnxs
3. "Architecture," Taj Mahal Official Website, http://www.tajmahal.gov.in/architecture.html
4. Barry Stoner (producer), *Treasures of the World: Taj Mahal,* "Building the Taj Mahal," 1999. http://www.pbs.org/treasuresoftheworld/a_nav/taj_nav/tajnav_level_1/3building_tajfrm.html

Chapter 4. Garden of Paradise

1. "Taj Garden," Taj Mahal Official Website, http://www.tajmahal.gov.in/taj_garden.html#waterwork
2. "Are These Ruins Part of the Rumored Second Taj Mahal?" *Smithsonian*, 2012, https://www.youtube.com/watch?v=qE1ZBAFURMU&nohtml5=False

Chapter 5. The Future of the Taj Mahal

1. "Taj Mahal" World Heritage Convention, http://whc.unesco.org/en/list/252
2. "The Taj Mahal Story," Taj Mahal Official Website, http://www.tajmahal.gov.in/taj_story.html
3. Marissa Fessenden, "The Taj Mahal Gardens Have a Special Relationship to the Solstice," *Smithsonian,* February 23, 2015. http://www.smithsonianmag.com/ist/?next=/smart-news/taj-mahal-gardens-have-special-solstice-relationship-180954136/
4. "Taj Mahal" World Heritage Convention, http://whc.unesco.org/en/list/252.

Works Consulted

Fessenden, Marissa. "The Taj Mahal Gardens Have a Special Relationship to the Solstice." *Smithsonian Magazine*, February 3, 2015. http://www.smithsonianmag.com/ist/?next=/smart-news/taj-mahal-gardens-have-special-solstice-relationship-180954136/

Great Buildings, *Architecture Week.* http://www.greatbuildings.com/buildings/Taj_Mahal.html

Jarus, Owen. "Taj Mahal Gardens Found to Align with the Solstice Sun." *Live Science,* February 2, 2015. http://www.livescience.com/49660-taj-mahal-gardens-align-solstice-sun.html

Official Site of the Taj Mahal. http://www.tajmahal.gov.in/

Osman, Jheni. *The World's Great Wonders—How They Were Made and Why They Are Amazing.* Melbourne: Lonely Planet, 2014.

Preston, Diana and Michael. *Taj Mahal: Passion and Genius at the Heart of the Moghul Empire.* New York: Walker & Co., 2007.

Taj Mahal: Memorial to Love. PBS Treasures of the World. http://www.pbs.org/treasuresoftheworld/a_nav/taj_nav/main_tajfrm.html

Taj Mahal. National Geographic Education. http://education.nationalgeographic.com/education/media/taj-basics/?ar_a=1

Taj Mahal. UNESCO World Heritage Site. http://whc.unesco.org/en/list/252

On the Internet

Deconstructing History: Taj Mahal. The History Channel.

http://www.history.com/topics/taj-mahal/videos

Engineering the Taj Mahal. The History Channel.

http://www.history.com/topics/taj-mahal/videos

Taj Mahal. Steven Carroll. History Channel, 2007.

aqueduct (AK-wuh-duhkt)—A bridge-like structure that carries water from one place to another.

artisan (AR-tih-sun)—Someone who is skilled at working with his or her hands at a particular craft.

axis (AK-siss)—The imaginary line through the middle of a shape.

calligraphy (kuh-LIG-ruh-fee)—The art of handwriting.

cenotaph (SEN-uh-taf)—An empty monument erected in memory of a deceased person.

dynasty (DY-nuh-stee)—A series of rulers belonging to the same family.

Hinduism (HIN-doo-iz-uhm)—One of the main religions of India.

inlay (IN-lay)—To insert or apply fine materials in the surface of an object.

irrigate (EER-ih-gayt)—To water a garden or field using pipes or other tools.

Islam (IZ-lahm)—A religion based on the words and philosophy of the prophet Mohammed, as written in the holy book, the Quran.

kiosk (KEE-osk)—A structure with open sides.

mausoleum (maw-zuh-LEE-um)—A building that houses a tomb or burial site.

minaret (MIH-nuh-ret)—A tall narrow tower used to call Muslim worshipers to prayer.

modification (mah-dih-fih-KAY-shun)—A change made in order to correct a problem.

mosque (MOSK)—A place of worship in the Muslim or Islamic religion.

Mughal (MOO-gul)—Persian descendants of Genghis Khan.

Muslim (MUZ-lum)—Someone who follows the religion of Islam.

patron (PAY-trin)—A person who gives money to support a project, artist, or organization.

pavilion (puh-VIL-yun)—An open-sided shelter often found in parks.

Persian (PUR-zhun)—A person from the ancient country of Persia, which is now Iran.

plinth (PLINTH)—A slab that is used as a base.

Quran (koo-RAHN)—The holy book of Islam; also spelled *Koran* or *Qu'ran*.

sandstone (SAND-stohn)—A kind of crumbly rock made from grains of sand stuck together.

PHOTO CREDITS: pp. 1, 25—Henrik Bennetson; p. 4—Public Domain; p. 8—Muhammad Karun; p. 10—Donelson; p. 11—Meredith P.; p. 12—Steve Evans; p. 14—Tarun Kamar; p. 16—Biswarup Ganguly; p. 18—Leon Yaakov; p. 20—Somsubhra Sharang; p. 21—Antoine Taveneaux; p. 27—Nedim Charbene. All other photos—Creatice Commons. Every measure has been taken to find all copyright holders of material used in this book. In the event any mistakes or omissions have happened within, attempts to correct them will be made in future editions of the book.

Index